ISBN 978-1-5284-0661-1
PIBN 10934735

1 MONTH OF
FREE
READING

at

www.ForgottenBooks.com

By purchasing this book you are eligible for one month membership to ForgottenBooks.com, giving you unlimited access to our entire collection of over 1,000,000 titles via our web site and mobile apps.

To claim your free month visit:
www.forgottenbooks.com/free934735

Historic, archived document

Do not assume content reflects current
scientific knowledge, policies, or practices.

United States Department of Agriculture,

OFFICE OF EXPERIMENT STATIONS—Circular No. 58.

A. C. TRUE, Director.

IRRIGATION IN THE VALLEY OF LOST RIVER, IDAHO.

By ALBERT EUGENE WRIGHT, *Agent and Expert.*

The valley of Lost River, Idaho, offers exceptional opportunities for interesting and instructive investigation. The annual flow of the river is very large, between 100,000 and 200,000 acre-feet of water. While no other river of its size sinks in its bed and rises again to the surface so many times, a study of the effect of heavy sinkage on the irrigation practices and water rights on Lost River will be applicable to like conditions on a large number of streams in Idaho and other arid States. Owing to the fact that the field work was begun too late to learn the use of water during June and July, this report of necessity deals with the duty of water somewhat in general. Changes in the surface flow of Lost River and seepage from the ditches are discussed more in detail, in so far as the data have a bearing on the distribution and use of water.

REGION STUDIED AND SOURCES OF WATER SUPPLY.

Lost River rises in the eastern part of Custer County, and in high water flows for about 60 miles toward the Snake River or until it reaches the edge of the Snake River Desert. There the channel, which has for some years been dry, turns north and may be traced for 40 miles to the point usually marked "Sinks" on maps of Idaho. Although the watershed contains 2,000 square miles, most of the water in Lost River flows from a district of high and very rugged mountains about 12 or 15 miles square. There is little timber or underbrush in these hills to retard the run-off, and the spring flood is sudden and heavy, reaching at times a discharge of 10 cubic feet per second for each square mile in this area. The fact that the flood subsides gradually and that the flow through the summer is well sustained may be due to the lodgment of snow in deep canyons and the percolation of most of the water through the deep beds of slide rock which form many of the mountain sides. The several forks at the head of the river join in one stream before reaching open and irrigable country. Below the junction the valley opens out gradually and the benches, formed by broad alluvial fans spreading out below the mouths of the canyons, become less steep.

Bed rock does not appear in the entire bed of the river, which has everywhere a gravelly bottom. Within a few miles of the point where the upper forks join, about 6 miles above Kinnickinnick Point, the water begins to sink into the gravel, which is of great depth and very clean and uniform. All the wells near this part of the river have encountered the same sort of material as far down as they have been dug.

Three or 4 miles below Kinnickinnick Point, or about in the center of the valley, which here widens out 6 or 8 miles, the channel is totally dry except during flood season. The dry section of the channel, which of course varies in length with the season, is known as the Big Sink or the Big Sinks. North of the sink Thousand Spring Valley spreads out, a tract 8 miles long and 2 or 3 wide, remarkably level, and with a slope so gentle that the slough which runs from it into Lost River flows with a scarcely perceptible current. On the south side, below Kinnickinnick Point, a gravelly bench slopes gently to the hills. On this bench, nearly back to the foothills, spring up two fine streams of water. They unite to form Warm Spring Creek, which flows swiftly over the top of the gravel in a shallow bed, discharging into Lost River the year round a steady flow of nearly 50 cubic feet per second. This stream and the slough flowing from Thousand Spring Valley, which does not probably average 10 cubic feet per second, are the only tributaries which normally reach the river in its whole length. For although a snow-fed stream may be found in nearly every canyon leading toward Lost River, without exception the water sinks almost as soon as it reaches the alluvial fan at the mouth of the canyon. It is only in extreme high water that these streams reach the river.

Above the point where Warm Spring Creek and Thousand Spring Slough come in, the water begins to reappear in the bed of Lost River. The location of the head of the new flow is as unsettled as the upper limit of the dry channel, varying with the stage of water. From this point the valley grows narrower, while the river bed itself widens out, forming a continuous stretch of wild-hay bottoms 10 miles in length and about a mile wide. The river has many channels and increases in volume down to a point called the Narrows, where high cliffs come in close on each side. Though bed rock does not appear in the channel at this point, the rising of the water is an indication that some impervious stratum lies near the surface here. This narrow place in the valley where the water becomes available again for irrigation divides the valley naturally into two sections, for most of the conditions governing irrigation practice are very different above and below this point.

For 30 miles below the Narrows the low benches on either side of the river have been taken up, forming an irrigated strip from 1 to

4 miles wide. Thirty-seven ditches bring about 30,000 acres of land under water, of which 12,900 acres were cultivated in 1903. The river flows on top of the ground for 12 miles below the Narrows, increasing slightly from springs in its bed. It then again begins to sink away very rapidly, and below Darlington the flow reaches a minimum.

Immediately it begins to increase from springs which come in from various sloughs and creeks until two-thirds of the former volume has reappeared. The ditches taken out from this part of the river have usually sufficient water. For 3 miles the water flows in the channel and then sinks rapidly until it is all gone. The location of this lower sink varies greatly, for in wet seasons the water has been known to flow from 12 to 20 miles down the channel, or to the edge of the desert, pretty well through the season. Nine ditches head on this lower part of the river, which in 1903 was dry, except for two or three weeks in June.

The valley of Lost River lies nearly 6,000 feet above sea level. The climate is very dry, the mean annual precipitation for several years at the ranch of Mr. Henry Harger being 8.41 inches. The heaviest rainfall for any month is in May, the average being 1.21 inches. The average total rainfall for June, July, and August is 1.56 inches. The mean annual temperature at the same station is 40.9° and for the summer months 62.8°.

LOSSES AND GAINS IN LOST RIVER.

ABOVE THE NARROWS.

The following measurements were taken to determine the gains and losses in the river above the Narrows and the effect of irrigation on the flow at the Narrows. The flow had just ceased to cross the Big Sink on August 7.

Sinkage measurements on Lost River above the Narrows, August 7–11, 1903.

Station where measured.	Distance from last station.	Discharge at station.	Diverted (−) or tributary (+).	Loss (−) or gain (+).	Loss (−) or gain (+) per mile.
	Miles.	*Cubic feet per second.*	*Cubic feet per second.*	*Cubic feet per second.*	*Cubic feet per second.*
Three miles above Uehren ranch	0	215.1	....	....	....
Uehren ranch	3	195.8	− 6.9	− 12.4	− 4.1
Kinnickinnick Point	5	101.1	−41.4	− 50.3	−10.1
One mile below Kinnickinnick Point	1	79.3	0.0	− 24.8	−24.8
Three miles below Kinnickinnick Point	2	56.9	0.0	− 22.4	−11.2
Five miles below Kinnickinnick Point	2	0.0	0.0	− 56.9	−28.5
Bridge at Battle Ground	6	145.5	a+25.0	+120.5	+20.1
Gage rod	1	163.0	0.0	+ 17.5	+17.5
Below Sharpe headgate	½	163.0	−11.3	+ 11.3	+22.6

a Of the 45 cubic feet per second flowing at the head of Warm Spring Creek 25 was diverted for irrigation, leaving 20 flowing into Lost River; 5 cubic feet per second is the estimated addition from Thousand Spring Slough.

The river fell an inch on the gage rod between August 7 and 11. So on the 7th, using the gage height as a basis, there was probably flowing out at the Narrows at the lowest station in the above table about 184 cubic feet per second instead of 163. Taking the basin above the Narrows as a whole, then, the conditions may be summarized as follows:

	Cubic feet per second.
Flowing in from all sources August 7:	
Lost River	215
Warm Springs	45
Thousand Springs	5
Total	265
Used for irrigation	85
Balance	180
Flowing out at the Narrows	184
Gain in the basin	4

The apparent indication is that 81 of the 85 cubic feet per second used for irrigation became unavailable for further use, only 4 cubic feet per second being returned to the river. But compare the foregoing results with the measurements given in the following table, which were taken August 26–29, two weeks after the water ceased to cross the sink and ten days after the diversions for irrigation had been reduced from 85 to 2 cubic feet per second.

Sinkage measurements on Lost River above the Narrows, August 26–29, 1903.

Station where measured.	Distance from last station.	Discharge at station.	Diverted (−) or tributary (+).	Loss (−) or gain (+).	Loss (−) or gain (+) per mile.
	Miles.	*Cubic feet per second.*	*Cubic feet per second.*	*Cubic feet per second.*	*Cubic feet per second.*
Eight miles above Kinnickinnick Point	0	a145.0			
Kinnickinnick Point	8	79.1	− 2.0	− 63.9	− 8.0
Two miles lower	2	47.9	0.0	− 31.2	−15.6
Two miles lower	2	0.0	0.0	− 47.9	−24.0
Gage rod	8	151.4	+50.0	+101.4	+12.7
Sharpe headgate	½	a162.4	0.0	+ 11.0	+22.0

a Estimated. These two discharges are based on other measurements which show that the loss between the upper station and Kinnickinnick Point and the gain between the gage rod and the lower station are very nearly constant quantities.

Again summing up the measurements in the basin as a whole:

	Cubic feet per second.
Flowing in from all sources August 26:	
Lost River	145
Warm Springs	45
Thousand Springs	5
Total	195
Used for irrigation	2
Balance	193
Flowing out at the Narrows	162
Loss in the basin	31

The apparent indication from this set of measurements is that between August 7 and August 26 the loss in the basin increased so that instead of a gain of 4 cubic feet per second there was a loss of 31 cubic feet per second. The natural expectation is that later in the season a still greater loss will be observed. But a set of measurements taken September 8 show facts quite the contrary:

	Cubic feet per second.
Flowing in from all sources September 8:	
Lost River	104
Warm Springs	45
Thousand Springs	5
Total	154
Used for irrigation	20
Balance	134
Flowing out at the Narrows	156
Gain in the basin	22

On August 7 there was a gain of 4 cubic feet per second; on August 26, a loss of 31; and on September 8, a gain of 22. Leaving these discrepancies for a moment, the question may be approached from another side. It is a fact observed by all the ranchers that the height of water in their wells depends on the stage of the river. Wells near the river rise and fall simultaneously with high and low water. Those farther back from the river rise and fall a week or two later than the corresponding changes in the flow of the river. When the spring flood comes the water continues for two or three weeks to disappear in the Big Sink, while the water plane slowly rises in wells in all parts of the basin. During the fall and winter the water plane slowly falls, reaching its lowest stage in the early spring. Of course it must not be supposed that this water plane is on a dead level. The river bed has a slope toward the Narrows of 15 or 20 feet to the mile, which, while it causes a very swift current in a river, may cause only a slow percolation through gravel.

Since this basin is a huge lake filled with gravel, it is obvious that, disregarding evaporation, the water diverted for irrigation must all find its way into the same bed of wet gravel as does the water that is not diverted from the channel and disappears in the sink. That is to say, it is no more lost when used for irrigation than when it goes into the sink. It is fair, then, to disregard the diversions for irrigation in computing the inflow and outflow of this basin.

The three sets of measurements may be summarized thus:

	August 7.	August 26.	September 8.
	Cubic feet per second.	Cubic feet. per second.	Cubic feet per second.
Flowing in	265	195	154
Flowing out	184	162	156
Gain (+) or loss (−)	− 81	− 33	+ 2

It is seen that while on August 7 the inflow was 81 cubic feet per second in excess of the flow at the Narrows, the two become more and more nearly equal till by September 8 the balance is in favor of the outflow. Thereafter the flow at the Narrows remains nearly constant, while the supply coming in above diminishes and remains very much smaller until the spring flood again reverses the ratio.

On August 17, 1903, the amount of water diverted above the Narrows was cut down to 2 or 3 cubic feet per second, for the benefit of older rights down the river. While the 60 cubic feet per second turned back into the main river made the water advance something like 100 feet farther across the Big Sink, this quantity of water, together with the 25 cubic feet per second turned down Warm Spring Creek, caused no noticeable increase in the flow at the Narrows. In fact, a series of gage-rod readings shows a maximum variation of only 0.02 foot, or one-fourth inch, from August 13 to the close of the field work on October 13. This represents a variation of but 7 cubic feet per second.

While the above measurements do not cover a period of time long enough to make them safely applicable to the use of water in all seasons, still the simplest inference from them is that it makes little difference whether the water flows down the river channel into the Big Sink and up again at the Narrows, or is diverted above the sink, carried across the bench, and allowed to sink into the ground among the roots of alfalfa and potatoes, to reach in time the same great bed of wet gravel.

One possible source of error in the above summaries lies in the uncertainty as to the source of the two springs which supply Warm Spring Creek. While the general topography shows no such indication, it is believed by some that the water that rises here is the same as that which sinks in the river bed above. For the quantity of water lost from the river in the 5 miles above Kinnickinnick Point was 50 cubic feet per second on August 8 and 55 on September 8, although the quantity at the upper station was meantime reduced one-half, which shows this loss to be fairly constant. On July 27 the Warm Springs discharged 59 cubic feet per second and on August 28, 47 cubic feet per second. The difference in level between the springs and the section of the river where the loss occurs, judging from the line of a ditch run on a grade of one-half inch to the rod, is at least 60 feet and possibly 100.

There is one other indication, which is more than a coincidence, suggesting that Warm Spring Creek carries a part of the same water measured at the upper station. The figures given in the three summaries show that the flood water which flowed across the sink from some time in June till August 7 failed to fill up the gravel bed to saturation (which is hard to believe), and that another month was required to make the inflow and discharge equal. But if we subtract from the supply as given in the summaries the 45 cubic feet per second rising in the Warm Springs, it is found that the water flowing at the Narrows balances the supply three or four weeks earlier, or at about the time when the flood ceases to cross the sink.

It should be noticed that the data given are insufficient to prove that all the water flowing into the basin appears again at the Narrows. It remains to be demonstrated that this basin has no other outlet. A complete record of the visible inflow and outflow for twelve months would go far toward settling the question, but allowances for evaporation and sources of underground loss or gain would still be conjectural.

To distribute water according to decreed priorities requires but small exercise of intelligence. But to distribute the water of Lost River with economy and fairness, either according to or in defiance of the decree, is impossible without a further knowledge of the effect of irrigation on certain land on the flow of the river in some distant part of the channel. Should the ranchers above the Narrows enjoin the water master from closing their gates it would hardly be possible for the holders of older rights below to prove actual damage from evidence which can at present be procured. On the other hand, should the holders of rights below enjoin the users above the Narrows from taking water, it would be equally impossible with present knowledge for the upper ranchers to prove that their use of water was not detrimental to users below. The best authority on the whole subject would be a set of gaging records, kept for at least twelve months, of the flow above the Uehren ranch and the flow at the Narrows. It is obvious that some conclusion regarding the source of Warm Spring Creek must first be reached, as its annual flow is not far from 30,000 acre-feet, or perhaps one-fourth of the total flow of Lost River.

BELOW THE NARROWS.

The following measurements were taken August 1–5 to find the losses from the river below the Narrows:

Sinkage measurements on Lost River below the Narrows.

Station where measured.	Distance from last station.	Discharge at station.	Diverted for irrigation.	Loss (−) or gain (+).	Loss (−) or gain (+) per mile.
	Miles.	*Cubic feet per second.*	*Cubic feet per second.*	*Cubic feet per second.*	*Cubic feet per second.*
Gage rod	0	215.3	0.0		
Smelter	3	201.7	20.6	+ 7.0	+ 2.3
Houston bridge	3	149.5	47.6	− 4.7	− 1.6
Gallagher ranch	5	155.1	10.6	+16.2	+ 3.2
Knapp ranch	1	145.1	12.1	+ 2.1	+ 2.1
Gra't ranch above Darlington	2½	57.1	14.5	−73.5	−29.4
Hanrahan ranch below Darlington	2	41.3	0.0	−15.8	− 7.9
Crossing below A. N. Anderson's	4½	30.0	60.4	+49.1	a +32.7
One-half mile below Moore	2	0.0	0.0	−30.0	−15.0

a The increase occurs in the first mile and a half below the Hanrahan ranch.

It is seen that from the Narrows down to the rocky point above Darlington, near the Knapp ranch, there is a pretty steady gain. It is not possible to determine whether all of the gain of 21 cubic feet per second is return water from irrigation or whether all or part is percolation from the mountain streams that sink before reaching the river. The discharges of Cedar, White Knob, Willow, and Alder creeks, which flow toward this section, aggregated about 25 cubic feet per second at the time of the measurements.

The important fact shown by the above gagings is the very serious loss of about 40 cubic feet per second of the 89 that sinks near Darlington. It would be an error to take this figure as the measure of a constant loss, for the amount of water which leaks away through a certain section of river bed is much greater when the water plane is low in the gravel and less when the gravel is saturated. In the basin above the Narrows, while one watches the water sink, he at least has the satisfaction of seeing the water level rise in his well and of knowing that the lost water will ultimately appear again at the Narrows. But this loss at Darlington of 40 cubic feet per second is a dead loss as far as availability for open-ditch irrigation is concerned, for it is evident, since not all the water that sinks reappears, that the water plane in the gravel requires a constant supply to keep it up to normal height. The unusually low level of water in wells and the drying out of the subsoil to a great depth following the dry season of 1892 show that there is constant leakage from the underlying gravel, for in the absence of supply there is a gradual lowering of the water level. To estimate the extent of this constant leakage is difficult without a study of the conditions for more than one season. To the shortage of 40 cubic feet per second in the water that rises below the Darlington sink must probably be added 20 cubic feet per second, which sinks in the bed of Antelope Creek opposite this point, a few cubic feet per second from Pass Creek and two or three small springs, and an unknown amount for seepage from ditches, laterals, and irrigated fields. The normal extent of leakage from this sink, then, is probably from 60 to 100 cubic feet per second.

The additional loss of 30 cubic feet per second, shown at the lowest station in the above table, was also a dead loss to the irrigators on Lost River. As it was the result of an attempt to force water down the channel to the ditches below, it will be discussed along with the distribution of water and the administration of water rights.

THE USE OF WATER ABOVE THE NARROWS.

In view of an unsuccessful attempt on the part of the ranchers above the Narrows to be placed in a separate district and to have their rights adjudicated as if the river above the Big Sink were an independent stream, the use of water by them has an interest to the whole valley out of proportion to the area they irrigate. For this reason the discussion of seepage and the duty of water will be given in some detail.

While three-quarters of the area now watered is in wild hay, this crop is mainly dependent on flooding by high water, the methods of irrigation being most primitive. The future expansion of irrigation must be on the bench lands, where varied crops are grown. The irrigation of the benches is accomplished by 10 or 12 ditches from the river and nearly as many from Warm Spring Creek.

SEEPAGE.

In this part of the valley, since the gravel is near the surface, the seepage is very heavy. The data will be most clearly shown in tabular form:

Seepage from ditches above the Narrows.

Ditch.	Length.	Amount diverted from river.	Amount used on land.	Loss by seepage.	Loss per mile.	Loss.	Remarks.
	Miles.	*Cubic ft. per sec.*	*Cubic ft. per sec.*	*Cubic ft. per sec.*	*Cubic ft. per sec.*	*Per ct.*	
Howell and Swendson.....	4	6.9	5.2	1.7	0.4	25	Grade excessive.
Frank Uehren............	½	1.6	1.3	.3	.6	19	
Bradshaw................	3	5.5	2.7	2.8	.9	51	Grade very excessive: ditch spreads out in places.
Wagon Tongue	a 6	21.8	13.7	8.1	1.2	37	Grade excessive; runs over clean gravel.
Davidson.................	12	7.4	2.8	4.6	.4	62	Grade moderate.
Mower Ditch from Warm Spring Creek.	3	4.1	3.4	.8	.3	20	Grade excessive.

a Length includes 7 laterals down to the point where measured.

It is noticed that all the ditches measured, except the Davidson Ditch, have excessive grades. The heavy loss in the Davidson Ditch was found to take place in the first mile, or before the ditch had "climbed" out on to the bench. As the river itself loses 50 cubic feet per second above the head of this ditch, it is to be expected that the ditch will lose heavily from that portion of it which runs through the coarse cobblestones of the river bottom. If the water should be measured at the present headgate the ditch would probably have to be abandoned, for 38 per cent of the legal allowance for 80 acres (1.6 cubic feet per second) would scarcely reach the ranch at all. But that part of the ditch which runs over the bench showed a loss of only 1.3 cubic feet per second in 10 or 11 miles, which is by far the lightest loss per mile of any of the ditches measured. This small loss is undoubtedly due to the moderate grade on which the ditch is run. The expense of puddling or fluming the section in which the heavy loss occurs would be nearly prohibitive.

The Wagon Tongue Ditch was built in 1900 by enlarging a very small ditch to the required capacity. This method saved the trouble of running any levels, but for so large a ditch the grade was of course very heavy. It carries water from the river to a large body of land halfway up Thousand Spring Valley, about 310 acres of which it waters. The swift current in this ditch can not gully because of large cobblestones in the subsoil, so the result is a swift stream 8 to 12 feet wide and half a foot deep, offering a very large surface for evaporation. The arrangement of laterals is also wasteful. The main ditch forks into seven little streams, of which the typical dimensions are 4 feet in width and 4 inches in depth. Were the seven laterals placed side by side they would make a stream 27 feet wide. This stream would be nearly a mile long and would have an average depth of less than 5 inches. Part of this bad construction could be avoided by abandoning parallel laterals and making the rest deeper.

The reason for the heavy grades on all the ditches from Lost River is due not to excessive fall in the valley, but to the necessity of heading the ditches at points where the supply in the river is reliable, that is, above the places of heavy sinkage. To avoid excessive grade a system of drops would be necessary in nearly every ditch.

DUTY OF WATER.

The situation above the Narrows is peculiar, and it is by no means improbable that none of the seepage water is lost. If this can be demonstrated it is obviously useless to take any precautions at all to reduce seepage, and if that is the case the best way for the ranchers under these ditches to save water is by preventing evaporation as far as they can. It is possible that deeper and narrower ditches, while increasing seepage, would so reduce evaporation as to make a real saving of water. Rotation between different laterals is not practiced at present. Few of the irrigators have yet been known voluntarily to shut off the water from their fields for even a short time, and they scarcely recognize any distinction between successive irrigations. But when the land has been more thoroughly worked and has become richer in humus the duty of water should be nearly doubled. Then, when the water shall be spread over the fields during only half as much time, evaporation can be reduced to a minimum.

The theoretical duty of water on Lost River is an "inch to the acre," or 50 acres for each cubic foot per second. A continuous flow of course is indicated. The season averages about a hundred days, beginning with high water early in June. As records have been kept only since July, 1903, the estimate of the actual duty of water is an approximation only. In the following tables that approximation is reached by taking it for granted that a measurement taken in the middle of the season is a fair average for the season. The data given are intended to be illustrative rather than exhaustive. For this purpose loss by seepage is eliminated, and only the water actually applied is taken into consideration. The "depth in season" is the depth to which the irrigating stream as measured would cover

the given area in 100 days. The fact that in 1903 most of the ditches
were shut off by the water master on August 17 is ignored.

Duty of water taken from Lost River above the Narrows.

Ditches.	Crops.	Area irrigated.	Water applied.			Remarks.
			Cubic feet per second.	Miner's inches per acre.	Depth in season, 100 days.	
		Acres.			*Feet.*	
Howell..........	Alfalfa......	14	1.2	4.3	17.2	Not noticeably over-irrigated.
Smith, Smith, and West.	Timothy.... Alfalfa......	50 90	9.9	3.5	14.0	Do.
Wagon Tongue.	Alfalfa and grain.	310	13.7	2.2	8.8	Soil more clayey.
Davidson........	Alfalfa...... Grain.......	25 50	2.8	1.9	a 1.9	

a Ditch used only 25 days in the year. Including the water from Navarre Creek which is applied to this same land, the depth of water used in the season is between 3.5 and 4 feet.

The first two ditches in the table water four ranches lying directly
on the river above Kinnickinnick Point, where the soil is especially
porous.   While 17.2 feet and 14 feet are not to be taken as accurate
measurements of a season's use of water, they show that the duty is
very low.   And apparently this heavy use of water is really needed
to raise crops on this land.   Mr. Frank Uehren, whose ranch lies
above Kinnickinnick Point, states that two rows of alfalfa which
accidently missed one of the five irrigations applied to the field were
almost entirely killed out as a result.   •
Under the Wagon Tongue Ditch .the average duty is seen to be
distinctly higher.   On the Joseph Clark ranch, which is the highest
land under this ditch, a stream of 1 cubic foot per second was used
on 60 acres, the duty being 0.87 inch to the acre.   The amount of
water applied to the 75 acres of bench land on the Davidson ranch is
pretty well determined, and is about the equivalent of an inch to the
acre for 100 days.
In the following table is given the quantity diverted per acre from
Warm Spring Creek for use on various branches.   As they lie near
the creek the measurements were made at the heads of the ditches.
The season is said to be "from along in April till the first of Decem-
ber," or 220 days.   One rancher accounts for the long run of water by
the fact that "without any headgates it is hard to shut it off."   There
is good reason to believe that the quantity diverted varies little during the season from the amount indicated.   There are logs laid in the
creek bed to form a sort of wing dam, which could not have stayed
in place with a rise of 3 inches in the creek.   The depth of water used
is the depth to which the stream of water measured would have covered the land irrigated by it if applied for 220 days; it can not be
taken as an accurate measure.

Duty of water taken from Warm Spring Creek.

Ditches.	Crops.	Area irrigated.	Water diverted.			Remarks.
			Cubic feet per second.	Miner's inches per acre.	Depth in season, 220 days.	
		Acres.			*Feet.*	
Henry Larter.......	Timothy...... Grain......... Alfalfa........	128 22 8	11.9	3.8	32.8	{ "Irrigates all { the time."
H. D. Baker.......	Alfalfa and timothy.	40	7.4	9.3	81.4	Crops do not look drowned.
A. B. Clark.........	Timothy......	64	5.6	4.4	38.4	
Mower Ditch	Grain......... Timothy......	35 25	4.1	3.4	28.7	}Season 210 days.

The table shows that each acre of the land in question is capable of absorbing indefinitely a steady flow of from 3½ to more than 9 miner's inches of water. In other words, the land is capable of absorbing a layer of water 1 foot deep in from three to seven days without swamping. It is obvious that the irrigation of land that really requires anything like the quantity of water indicated above would be an absurd waste were it not probable that the excess used returns to the river at the Narrows. So confident are the people that this is the case that as yet no objection has been raised by irrigators below the Narrows. In 1903 the diversion of water from Warm Spring Creek was not regulated by the water master at all until he shut off the water altogether on August 17. In fact, every question as to the supply, waste, and use of water above the Narrows is seen to hinge directly on the question whether or not all the water entering the basin rises and flows out at the Narrows.

On the 5,602 acres of land actually irrigated above the Narrows about 4,000 is in wild hay. These hay bottoms are irrigated by high water, which fills numerous sloughs and creeks from which water is easily made to flood the land by crude dams. Very few attempts are made to put permanent diversion weirs across the main channel, as these lands lie very low along the river and the irrigator has only to assist the natural flooding of his field. In places the water rises and in others it sinks, making it well-nigh impossible to estimate the amount of water actually applied. Weirs and headgates are unknown and, in fact, are unnecessary, for usually the water is as far as possible turned off by July 15 (in 1903 on August 1) in order to dry the ground for cutting, and the supply is still abundant on that date. If it can be shown that the excess of water used in flooding these meadows either flows back into the river or, sinking into the ground, rises again at the Narrows, the quantity of water necessary to raise wild hay would be exactly the amount evaporated into the air added to the amount absorbed by the grass. It is probable, then, that the quantity of water consumed in raising a ton of hay is less on these meadows than in any other part of the valley. Here again the question appears to rest on the uncertainty as to the quantity of water returned to the river at the Narrows.

USE OF WATER BELOW THE NARROWS.

DIVERSIONS.

Below the Narrows are some 34 ditches taking water from Lost River. The following list is taken from the records of the water master who collected this material from the ranchers themselves as a basis for the distribution of water. As it was not generally understood when the data were being collected that the water would be delivered on the basis of "an inch to each acre irrigated," the acreage given is probably accurate within the limits of good guessing, except that where an entire forty was reported no allowance was made in any case for roads, fences, ditches, buildings, or for the river channel; and in some cases there is included in the acreage given as "grass" considerable areas more or less covered with willows, cottonwood, and quaking aspen. The ditches are numbered in order from the Narrows down.

Ditches taking water from Lost River below the Narrows and the acreage, crops, and claims under them.

No.	Name of ditch.	Length.	Alfalfa.	Grass.	Grain.	Garden truck.	Not reported.	Total.	Total acreage under ditch.	Unreclaimed land under ditch.	Amount of water decreed.
		Miles.	Acres.	Acres.	Acres.	Acres.	Acres.	Acres.	Acres.	Acres.	Miner's inches.
1	Sharpe	8	252	84	105	a 113	......	554	1,676	1,122	1,200
2	Blaes	4	144	52	205	10	......	411	618	207	650
3	Irish	7	348	236	140	25	......	749	988	239	1,014
4	Rogers and Miller	6	240	234	76	5	......	555	1,229	674	1,125
5	Vannons	2½	......	240	15	......	......	255	255	......	160
6	Burnett	12	249	170	360	18	......	797	1,721	924	1,655
7	D. P. Wells	½	20	55	40	......	......	115	350	235	350
8	Harris	4	145	145	155	6	......	451	739	288	675
9	East Side	4½	224	......	182	31	......	437	1,449	1,012	1,560
10	Eberhardt	3½	35	......	56	5	......	96	540	444	460
11	O'Neal, McKinney, and Donohue	5	215	......	100	30	......	345	640	295	640
12	Wells	4	138	......	35	3	4	180	440	260	340
13	Lower Burnett	5	240	130	60	......	......	430	480	50	480
14	Burnett and McGowan	4	120	......	50	......	60	230	760	530	960
15	Hendricks and Gra\	1½	45	20	......	......	......	65	320	255	320
16	Beers and Carlson	3	165	180	75	......	......	420	670	250	670
17	J. Rabido	1	90	......	......	2	......	92	160	68	160
18	Stickney	6	100	40	15	1	......	156	160	4	160
19	Hash	3½	190	40	●12	1	......	243	385	142	325
20	Hanrahan and Darlington	3	150	......	......	......	......	150	150	......	150
21	Burstedt and Johnson	3	124	......	14	3	......	141	560	419	560
22	West Side	9	649	220	131	21	72	1,093	2,440	1,347	2,476
23	Harger and Sullivan	11	803	......	70	29	120	1,022	2,840	1,818	2,855
24	Anderson and Darnle\	3	484	6	46	2	......	538	1,000	462	865
25	Island	10	1,154	486	147	21	......	1,808	4,360	2,552	4,360
26	Munse\	7	110	110	14	5	......	239	793	554	800
27	Ed. James (James Creek)	1½	60	......	......	......	......	60	60	......	60
28	A. Viso (James Creek)	2	115	......	25	4	......	144	640	496	580
29	M. M. Moore	2	14	20	......	......	......	34	160	126	216
30	Mark Hurst	4	......	160	......	......	......	160	160	......	160
31	Walker	2	202	50	8	2	......	262	400	138	400
32	Arco Trading Co. or George Ferris	4	160	40	......	......	......	200	960	760	1,000
33	Fleischer	3	86	60	45	3	......	194	360	166	480
34	Jones	2	138	10	50	4	......	202	674	472	760
35	Hahn	4	35	......	......	1	......	36	400	364	400
36	Paul Thomas	½	80	50	10	......	......	140	200	60	320
37	Ira George	1	......	......	......	......	200	200	200	......	200
	Total	......	7,324	2,838	2,241	a 345	456	13,204	29,937	16,733	29,546

a Includes 100 acres for the town of Macka\.

14

This table is given as a basis for the following discussions of duty and seepage. The ditches will be referred to by the numbers given in the table.

DUTY OF WATER.

With no data as to the use of water before July 1, when the water master reduced the diversions to "an inch to the acre for each acre actually irrigated," a fair approximation of the duty of water can hardly be reached. But in no case under the first 25 ditches in the above table is it likely that the amount used was less than the equivalent of an inch to the acre for 68 days, or a depth of 2½ feet, or more than an inch to the acre for 125 days, or 5 feet in depth. In the case of the ditches below the Island Ditch (No. 25) the supply failed in the river, and the water used was limited to a very short season.

On the ranch of Geo. E. Walker (Ditch No. 31) 200 acres of alfalfa received one good irrigation for 14 days in June, and yielded 300 tons, or one-half of a full crop.

On the ranch of Antone Viso (Ditch No. 28), where alfalfa was badly burned in 1902, one irrigation gave from ½ to 1 ton per acre with one cutting. A better supply of water was secured later in the season and the second cutting proved much heavier.

As is the case above the Narrows, the quantity of water that the land is capable of absorbing varies with the distance of the water plane below the surface. On the Island, which lies between the present river channel and a high-water channel, it is said that a flow of 100 inches, or 2 cubic feet per second, failed to saturate one-quarter of an acre of unbroken land in seven days in the late fall, although the water used would cover the area to a depth of 112 feet. On evidence of this sort, a claim was made that the duty of water on land under the Island Ditch should be fixed by court decree at 2½ inches to the acre. But this claim was disregarded.

In general, alfalfa receiving a continuous flow through the season of 1 inch to the acre showed no signs of over-irrigation. A large acreage in alfalfa had no water from August 7 to September 5. The result in many cases was a shortage in the second crop, and in all cases the high spots were more or less severely burned. Oats did not seem to suffer at all during the same dry month.

The first reason why alfalfa burns so quickly is that the subsoil is very porous. This can hardly be remedied, but the other causes—uneven ground, lack of humus in the soil, and unshaded bare spaces—can all be largely obviated. With ordinary crops repeated plowing and cultivation will level the surface in the course of a few years. But alfalfa land must be leveled before sowing if it is to be even enough for economical irrigation. The lack of vegetable mold or humus can be supplied only by years of cultivation and the rotation of crops. The thin stand of alfalfa must be remedied by heavier seeding and by frequent reseeding where thinnest. It is an error to hope to get or keep a good heavy stand of alfalfa with the light seeding practiced in lower altitudes.

The loss of water by seepage from the ditches was measured in several different parts of the valley. As most of the ditches have irrigated land above them, it is not possible to avoid an error due to waste and seepage running in from those lands. In several cases an actual gain was observed where one ditch ran close to another for some distance. As these errors are all on one side, the losses given represent the minimum losses to be expected.

The Sharpe Ditch (No. 1), carrying 19 cubic feet per second, lost 1 cubic foot per second in a mile, or 5 per cent. A small lateral which supplies the town of Mackay lost 0.2 cubic foot per second, or 12 per cent in one-half mile. In both cases the grade is excessive.

The Blaes Ditch (No. 2), carrying 12 cubic feet per second, lost 2.5 cubic feet per second, or 21 per cent in 1 mile, near the head, while the section for a mile and a half below the Blaes ranch, where it flows a few rods below the Sharpe Ditch, showed a gain of more than 1 cubic foot per second. In a 1-mile section lower down on the ditch where the grade is much lighter and the subsoil less gravelly, there was a loss of less than 0.3 cubic foot per second, or not quite 2½ per cent.

The loss from the Irish Ditch (No. 3) in 3 miles was only 3 per cent. There was undoubtedly some seepage into the ditch from irrigation going on above it.

The West Side Ditch (No. 22), carrying 18.7 cubic feet per second, lost 2.7 cubic feet per second, or 15 per cent in 3½ miles, an average loss of 0.8 cubic foot per second per mile.

The upper Harger Ditch (part of No. 23), carrying 5.8 cubic feet per second, lost 1.7 cubic feet per second, or 29 per cent in 4 miles, an average of 0.4 cubic foot per second per mile. In a 1-mile section below Howell's lateral, where the grade of the ditch was considerably increased to avoid a rocky knoll, there was a loss of 1 cubic foot per second, or 25 per cent. The total loss in 5 miles was 2.7 cubic feet per second, or 46 per cent. The ditch was not running more than one-fourth of its capacity, so that the percentages given would not hold good with a full ditch.

The lower Harger Ditch (part of No. 23), carrying 11.9 cubic feet per second, lost in the first mile 3.6 cubic feet per second, or 30 per cent. The loss in the next 3 miles was only 0.7 cubic foot per second, or 8 per cent, an average of 0.25 cubic foot per second per mile. The average loss for the 4 miles was 1.1 cubic feet per second per mile. The Harger ditches are run on pretty even grades, with good alignment, and except in the section of 1 mile just noted and that part of the upper ditch in which the fall is increased, the loss from seepage is not excessive.

The ditch connecting the upper and lower Harger Ditches is very steep and shows a loss of 1.7 cubic feet per second in 1 mile, or 14 per cent.

A small lateral near the lower end of the upper Harger Ditch, carrying 2.5 cubic feet per second, lost 0.2 cubic foot per second, or 8 per cent in 1 mile. The grade was very steep and the ditch had cut a deep channel in places.

The Darnley Ditch (No. 24), carrying 6.9 cubic feet per second, lost 5 per cent in one-quarter of a mile. The total loss in 4 miles was 2.4 cubic feet per second, or 35 per cent, the average loss per mile being 0.6 cubic foot per second.

The Island Ditch (No. 25), carrying 15.7 cubic feet per second, lost 0.5 cubic foot per second per mile for 5 miles, making a loss of 17 per cent. The lower end of the ditch lost much less, the seepage being only 0.2 cubic foot per second per mile, or 5 per cent in 3 miles.

Hooper's lateral from this ditch, carrying 1 cubic foot per second, lost 70 per cent in 6 miles. This lateral is an excellent example of the wastefulness of a long ditch having a very small capacity. While it lost only 0.11 cubic foot per second per mile, or about 6 miner's inches, that loss was so great a percentage of the total stream that in 6 miles more than two-thirds of the water was lost.

There are two reasons why small ditches lose more water in proportion than large ones. First, the loss varies with the area of the surface that is wet. Thus six ditches, each carrying 1 cubic foot per second, wet a much greater surface on sides and bottoms than one ditch carrying 6 cubic feet per second. The second reason is that seepage is nearly independent of pressure or the depth of water in the ditch.

This fact suggests the remedies for excessive seepage. First, the surface wet by the 157 miles of ditches below the Narrows can be reduced by abandoning part of the smaller ditches where they run parallel and deepening the rest so that they will carry all the water required. Where that is not feasible a good deal of water can be saved by rotating or "bunching" water as far as possible among the small ditches and among the users from the larger ones; the ditches and laterals need be wet only half the time. While such an arrangement can not be forced on a community it would be to the interest of all, and, with adequate measuring devices, it offers no greater likelihood of unfair distribution than a system of continuous flow.

The second cause of heavy seepage—the porous formation over which the ditches run—can be overcome only by placing finer material in each ditch as a lining. But before that can be done effectually, the grade must be reduced so that the current will not wash the fine material away. This can be accomplished only by a system of wooden drops or of paved rapids placed at intervals in the ditches. In most cases silt from the river will be found sufficient to stop the leakage to a large extent. In the Rogers and Miller Ditch (No. 4) clay was used as a puddle with good success. Either of these methods is expensive, and, though their efficacy is undoubted, it is not certain that the saving in water would warrant the expense. This can be found out only by trying.

THE DISTRIBUTION OF WATER FROM LOST RIVER.

In 1903 for the first time the water of Lost River was distributed by a public officer, according to a decree rendered by a district court in January of that year. The scarcity of water which precipitated the water suit in 1902 was due to two causes. While no official records of the snowfall have been kept, it is held by all the ranchers above the Narrows, and it is admitted by many below, that the annual snowfall in the hills averaged very much heavier in the eighties and early nineties than in any year since, and that the run-off has been correspondingly short. The other cause, which is loudly urged by the lower ranchers, is the settlement of the valley and the consequent increased diversion of water for irrigation.

Nearly 230 irrigators were parties to the suit, which resulted in the adjudication of 323 rights to water. In the absence of any reliable information regarding the nature and capacity of the various sinks on Lost River, the court ignored them completely, thus, while establishing a system of priorities, avoiding all responsibility for an economical distribution of water thereunder. The suit ended in a compromise. Each claimant was awarded an inch to the acre for all the land under ditch to which he held title, the right dating from the time of building his ditch.

The size of each ditch and the area of land under it was looked into in the most superficial manner. As is shown in the table of ditches given above, there are 10 ditches (Nos. 1, 4, 5, 6, 8, 10, 12, 19, 24, and 28) in the report, of which the number of acres under ditch exceeds the number of inches decreed. The excess aggregates 1,240 acres. There are 12 cases (Nos. 2, 3, 9, 14, 22, 23, 26, 29, 32, 33, 34, and 36), aggregating 849 inches, in which the number of inches decreed exceeds the acreage under ditch, as estimated by the ranchers themselves when questioned by the water master. In spite of the fact that between 5,000 and 6,000 acres of the land in question was natural hay meadow, unbroken and partially self-irrigated; in spite of the varying duty of water in different parts of the valley, and with no regard for the fact that three-fifths of the land in question was still unreclaimed sagebrush land on which water had never been used, the court complacently " ordered, adjudged, and decreed " as follows:[a]

(1) That each and every of the tracts of land described in the complaint, answers, and cross-complaints in this action as belonging to the said several parties plaintiff and defend-

_____

a The italics are not in the original.

ant herein are *arid in character* and will not successfully raise any agricultural crop witl out the use of water thereon.

(2) That *the quantity of water required for the successful irrigation and cultivation of sa: land is one inch per acre* measured under a four-inch pressure, or its equivalent in cubi feet per second of time, to be measured at the point or place of diversion.

(3) That each of the parties plaintiff and defendant hereinafter named were, at th times of the several appropriations of water made as hereinafter set forth, the owner respectively of the several tracts of land set forth and described in the complaints, answers and cross complaints, and that for the irrigation of the same the said parties or their prede cessors in interest did respectively appropriate the quantities of water hereinafter stated and on the dates respectively as hereinafter stated, and set opposite the names of each o said parties, such water being estimated and measured under a four-inch pressure or it: equivalent in cubic feet per second of time; *and ever since said date of the said severai appropriations said water has been so diverted and used on said lands in the irrigation thereof.*

The following-named persons, parties to this suit, are entitled to the use of the waters of Lost River and its tributaries upon the lands mentioned and described in the complaint and cross-complaints and answers herein in the following amounts and from the following dates, to wit:

[Here follows a list of the 230 claimants and their 323 awards.]

That each of the parties to said suit, plaintiff and defendant, and their successors in interest, their agents, attorneys, servants, and employees, and any and all persons acting in aid or for any of said parties, are hereby forever restrained from in any manner inter fering with or diverting or using the waters of said Lost River, except in accordance with the terms of this decree.

In the following table the number of inches opposite each date includes all the water rights dated in that year or earlier, so that each number represents the quantity of water required to satisfy all rights down to the date in question. The "other creeks and springs" are Alder, Beda, Cedar, Upper Cedar, Corral, Lehman, Navarre, Pass, Lower Pass, Sage, Willow, and Rock creeks, and Cristo, Condon, and Jensen Creek springs.

Summary of decreed water rights.

Date of priority as decreed.	From Lost River below Narrows (including James Creek).	From Lost River above Narrows.	From Warm Spring Creek.	Total from Lost River and actual tributaries.	From Antelope Creek.	From other creeks and springs.	Grand total.
	Miner's inches. a	*Miner's inches.*	*Miner's inches.*	*Miner's inches.*	*Miner's inches.*	*Miner's inches.*	*Miner's inches.*
1875	....	260	....	260	....	....	260
1879	....	260	....	260	410	30	700
1880	....	260	....	260	810	210	1,280
1881	....	435	....	435	810	210	1,455
1882	....	805	....	805	810	1,214	2,829
1883	886	805	....	1,691	810	1,344	3,845
1884	6,900	1,198	....	8,098	970	1,389	10,457
1885	10,450	1,278	....	11,728	1,900	1,749	15,377
1886	14,000	1,518	160	15,678	2,490	1,949	20,117
1887	16,522	2,078	160	18,760	2,430	1,949	23,199
1888	17,432	2,658	160	20,250	2,960	2,139	25,349
1889	18,837	2,658	160	21,655	2,960	2,299	26,914
1890	19,466	2,818	280	22,564	2,960	2,299	27,823
1891	20,206	3,128	280	23,614	2,960	2,299	28,873
1892	20,680	3,258	380	24,318	3,110	2,359	29,787
1893	20,680	3,711	380	24,771	3,110	2,519	30,400
1894	23,712	3,861	380	27,953	3,110	2,519	33,582
1895	24,972	4,021	380	29,373	3,110	2,519	35,002
1896	29,432	4,181	380	33,993	3,110	2,594	39,697
1897	32,812	4,441	380	37,633	3,270	2,594	43,497
1898	34,142	4,641	380	39,163	3,430	2,594	45,187
1899	36,009	5,121	380	41,579	3,430	2,594	47,594
1900	36,914	5,521	380	42,815	3,530	2,594	48,939
1901	37,684	5,976	380	44,040	3,530	2,594	50,164
1902	38,657	7,176	430	46,263	3,530	2,594	52,387

a The miner's inch is equivalent to one-fiftieth of a cubic foot per second by statute in Idaho.

In this summary from the printed decree it is noticed that rights to 38,657 miner's inches of water were granted to ditch owners below the Narrows, while in the table prepared by the water master the total of 29,546 miner's inches is given as decreed for the 37 ditches. Part of this discrepancy of over 9,000 inches is due to a decree of 2,500 inches to the Houston Ditch Company. The Houston Ditch (now in private ownership) was out of repair in 1903 and so is not listed with the rest. The remaining difference of 6,500 inches is caused by the present use or total abandonment of the land for which it was decreed, so that it did not come within the official province of the water master.

THE ADMINISTRATION OF DECREED RIGHTS IN 1903.

The distribution of water among the 37 ditches below the Narrows formed the principal part of the water master's work. The use of water from Lost River and Warm Spring Creek above the Narrows was not restricted at all till August 17, because the irrigation of hay bottoms had ceased before low water and because the benefit of such regulation to the ranchers below seemed very doubtful. Antelope Creek waters nearly 2,500 acres of wild hay meadows and less than 300 acres of other crops, so that no regulation whatever of the use of water from that stream was needed. The diversions from "other creeks and springs" were regulated in part, but since their flow in every case failed to reach the river, the regulation was not enforced, and it is very doubtful if the decree can be enforced against them in the future.

For convenience in distributing the water the rights granted from Lost River and its tributaries were divided into "priority classes," or blocks, lettered from A to F, each block containing rights to about 8,000 inches. Thus class "A" includes rights from 1875 to 1884, class "B" from 1885 to 1886, class "C" from 1887 to 1891, class "D" from 1892 to 1895, class "E" from 1896 to 1898, and class "F" from 1899 to 1902. When water is scarce the supply is cut off by classes, beginning with the last. For his own use the water master prepared also a table of ditches giving the number of inches of "A," "B," and "C" rights, and so forth, belonging to land under each ditch, and the number of acres actually irrigated under rights of each priority class.

When the water master began his work on July 1 it was expected that the reduction of diversions to an inch to the acre would assure everyone plenty of water, for at that date the flow of Lost River at the Narrows was about 450 cubic feet per second, or 22,500 miner's inches, which is equal to a flow of 1.7 inches for each acre irrigated by the 37 ditches. But in spite of the fact that the diversions were cut down to an inch for each acre irrigated, the water failed to reach the headgates of the lower ditches, and in fact the flow slowly retreated up the river channel.

By August 1 it became evident that the lower ditches would get no water unless more was turned down from above. In the nine lower ditches are rights to 10 inches in class "A" and 1,190 inches in class "B." But of the land represented by these rights a total of less than 300 acres is irrigated. On August 6 and 7 all the "F," "E," and "D" rights were cut off from the Narrows down, 2,500 miner's inches, or 50 cubic feet per second being turned back into the channel. This cut included all rights dating later than 1891. At the time of the cut the water was flowing in the channel for a short distance below Moore. The increase of 50 cubic feet per second caused it to advance about a quarter of a mile farther down the river bed.

On August 15 and 16 all "C" rights were cut off, 1,600 miner's inches, or 32 cubic feet per second, being turned back into the river. This left water for only the rights dating before 1887. This second cut caused a temporary advance of the water down the channel, and by August 18 the head of the water was moving down the channel at the rate of about 10 feet per hour. But by August 20 it had failed to reach more than half a mile below the Moore crossing. The entire reduction of 4,100 inches, or 82 cubic feet per second, failed to benefit the holders of "A" and "B" rights in the least, as the water did not come within 5 miles of the headgates. On the other hand, 4,100 acres of land was left dry for from 20 to 30 days, or until the departure of the water master on September 5 raised the embargo.

The injury to the second crop of alfalfa was in some cases severe. It is probable that 2,000 acres of alfalfa was injured to the extent of a quarter of a ton of hay to the acre, or a loss of at least $2,000. A thousand acres more lost probably half a ton to the acre, or say $2,000 more. The remaining area which was short of water owing to the cut was mostly in grass and grain, and was not badly injured. The ranchers stood this loss of at least $4,000 without a murmur in the hope that it would demonstrate the impossibility of forcing water across the lower sink in the natural river channel.

On August 17, in order to make the experiment complete, the water master shut down the ditches above the Narrows, leaving only 2 or 3 cubic feet per second for culinary use. As it happened only one or two holders of "A" and "B" rights were then using water, so the cut was nearly complete. Even the springs which never reach the river were turned down their natural channels. As has been seen (page 4) the effect on the flow at the Narrows was harly appreciable. The injury to crops was not great except to young alfalfa, which was somewhat burned in places.

The result of this attempt to administer the decree literally was not encouraging to the ranchers for whom the water was turned down. For five years the lower ranchers have been short of water. As we have seen, in 1903, when the supply was better than usual, it failed to reach them even when all late appropriations were cut off. The question of how to administer the decree with economy and justice is still to be solved.

PROBABLE SHORTAGE OF WATER IN THE FUTURE.

While in 1903 the supply of water in Lost River would have been ample for every acre irrigated from it below the Narrows, it is important to notice that the reclamation of the sagebrush land that is now under ditch is bound to work hardship to later appropriators. If all the land for which water is decreed had been actually cultivated in 1903 the distribution of water would have been as follows, granting that the flow at the gage rod represents the quantity of water available for irrigation: On July 1 the supply of 450 cubic feet per second furnished enough for all rights down to 1894, or for 22,500 acres. On July 15 the flow of 300 cubic feet per second would have supplied all rights down to 1886 and some of those dating from 1887, or for 15,000 acres in all. That is, only "A" and "B" rights and "C" rights to the extent of 1,000 inches would have had water at all after July 15. On August 1, 215 cubic feet per second would have satisfied rights down to 1885, giving 400 inches to 1886 rights, or would have supplied water for 10,750 acres. On August 15 the stream carried 156 cubic feet per second, sufficient to supply rights down to 1884 and give 900 inches to holders of 1885 rights.

This fact that only "A" and "B" rights would be of any value if all the decreed rights were used is hanging over the heads of all later appropriators who, of course, dare not improve their land lest earlier appropriators subsequently reclaim theirs and claim all the water. The new Idaho irrigation law is intended to cover this difficulty by a provision that—

All rights to the use of water acquired under this act or otherwise, shall be lost and abandoned by a failure, for the term of two years, to apply it to the beneficial use for which it was appropriated, and when any right to the use of water shall be lost through nonuse or abandonment, such right to such water shall revert to the State and be again subject to appropriation under this act.[a]

It is obvious that this two-year clause can not be called into action before 1905, as the finding in the decree is conclusive that all the water decreed had been in continuous use up to January, 1903.

THE VALUE OF WATER.

As the value of water is the fundamental consideration which must decide questions of storage and methods of conducting and applying water, as well as fix the price of land and index the real wealth of the community, it will be briefly discussed in conclusion. It is so intimately involved with questions of land, water supply, crops, markets,

---

[a] Laws of 1903, p. 234.

and the personnel of the ranchers that it must be approached by a consideration of general conditions. The land in the valley of Lost River is rich and capable of supporting any crop that can stand the climate. The water supply with present methods of applying it to the land is sufficient, in average seasons, for the successful irrigation of all land now reclaimed. Before discussing the possibility of doubling the irrigated area by storage, the crops grown must be considered.

The most important crop at present is alfalfa. It yields an average of 2 tons to the acre the first cutting, and from 1 to 1½ the second, making the average annual yield about 3 tons to the acre. At $4 per ton the crop produces $12 per acre. The cost of cutting and stacking is from 90 cents to $1.50 per ton, leaving a profit of from $7.50 to $9.30 per acre. Land is assessed at $4 an acre regardless of improvements, the average tax being about 16 cents an acre annually. The capital invested in an acre of alfalfa land is $1.25 for the desert land, say $2 for ditches, $6 for clearing sagebrush and plowing, and $3 for seed, making a total of $12.25. Allowing 10 per cent on this investment and 16 cents an acre for taxes, there remains a net profit of from $6.10 to $7.90 an acre. If 4 acre-feet of water is needed to produce this profit the value of water for the irrigation of alfalfa is from $1.50 to $2 an acre-foot. A miner's inch flowing 100 days, which is equivalent to 4 acre-feet, is therefore worth from $6 to $8.

Applying this value of water in a single instance, could the loss of 40 cubic feet per second in the Darlington Sink be prevented during the season of 100 days, the water saved could be made to yield a profit of from $12,000 to $16,000 a year.

But if the profits are so generous, why is it that the early appropriators, who have an undoubted claim to water, have left unreclaimed thousands of acres, content with half the profit the land might yield and cheerfully paying 16 to 20 cents an acre every year in taxes? The first reason is that without capital the reclamation of dry land means hard work. As one rancher put it, "You can't round up a crop of alfalfa on a cow pony." This difficulty is by no means new in cattle raising communities. The other reason is the lack of settlers. The ranches are too large for intensive farming. It is probable that with heavier seeding, careful leveling of the ground, and more painstaking irrigation the average crop of alfalfa can be increased to at least 4 tons to the acre. The conditions on Lost River are peculiar, and it is for the ranchers themselves to discover by experiment what careful farming will produce in the valley.

The high price of hay in the valley is kept up by the great demand for the winter feeding of cattle. For several years the price of wild hay has been $5 per ton. There are theoretically about 9,000 acres of wild hay bottoms in the valley, but of this area perhaps 1,500 acres is cumbered with willows, and of the remaining 7,500 acres not more than

5,000 has ever been cut. An acre of wild hay yields from 1 to 1¼ tons, worth from $5 to $6.25. It costs $1.25 per ton to cut and stack, leaving a profit of from $3.75 to $4.70 per acre. The stubble is worth 50 cents an acre for pasturing, raising the annual profit on 1 acre to $4.25 or $5.20. That part of the meadows which can not be cut on account of bushes or surface water is worth from $2.50 to $3.50 per acre as winter pasture. It is seen that a very fair profit can be got from a hay meadow with no work whatever, as the crop yields only about $1 less per acre as pasture than when cut.

The possibility of injuring the 4,000 acres of meadow lying above the Narrows by diverting the river around the Big Sink or by storing the flood water in Thousand Spring Valley would be an important item in determining the feasibility of either enterprise.

Some interesting experiments are to be tried in 1904 with sugar beets. The ideal climate for beets exists in that belt across the United States·where the temperature during June, July, and August averages 70°. South of this line the yield is heavy but deficient in sugar. North of this line the sugar content is greater but the yield is less reliable. At the station of the Weather Bureau at Lost River the summer temperature averages 62.8°. If beets can be made a sure crop, the high percentage of sugar would in a measure offset the severe climatic conditions. The other necessity for successful sugar-beet culture is a short haul to a factory. It is essential that there be at least 4,000 acres in beets within hauling distance. On Lost River the ease with which water power could be developed might be some inducement to a sugar company.

In view of the fact that so large an area having early water rights remains unreclaimed, it is not surprising to find very little attention paid to the raising of potatoes, garden truck, and small fruits. Potatoes are grown in half-acre lots in all parts of the valley, but as yet the home market has never been supplied. One dealer in Mackay shipped in 150 tons of potatoes in 1903, and another paid out $700 in 1902 for potatoes raised on the Snake River.

Even the local crop of oats is not nearly equal to the local demand. In 1902 nearly $10,000 was sent out for grain from Mackay alone. There is no good reason why the valley should not more than supply the growing home market, as oats can be raised on any of the bench land.

When well cared for small fruits flourish in the valley. Gooseberries and currants do exceptionally well. If they were laid down in the fall, raspberries and even blackberries would probably be profitable. Strawberries do fairly well and, since the crop is very late, the market is always good. On the more sheltered ranches hardy apples, plums, and pears grow nicely.

The crops which have been mentioned require a great deal of work, but in proportion to the amount of water required to raise them they yield ten times the values that can be made in alfalfa or grass or grain.

The valley of Lost River has an advantage in its location that goes far toward offsetting the disadvantage of high altitude. The town of Mackay, called into existence by the White Knob copper mine, furnishes a larger market for oats, potatoes, other vegetables, and small fruits than the valley has yet supplied. And this valley is the natural gateway for a very large mining area which can be utilized only for stock. A spur of the Oregon Short Line runs up through the valley as far as Mackay, and may some day be extended to Salmon. Lost River can provide a large and growing mining district with all food products except pork and the less hardy fruits, and it has the advantage of a 100-mile haul over Blackfoot on the Snake River. The magnificent range in the hills around Lost River is bound to be overstocked in time, and, when cattle yield less and less profit, it is the intensive cultivation of the soil that is going to pay.

As long as the ranchers put their trust in prospects and sink their resources in prospect holes, as long as they sit waiting for development, the valley will remain undeveloped. But if by more economical methods of use and distribution of water and by the cultivation of more valuable crops the value of water shall be doubled, it will be profitable to double also the available supply by storage, and Lost River will support twice its present population and produce four times its present wealth.

ACKNOWLEDGMENTS.

Especial acknowledgments are due to Mr. John W. Sheppard, who, as water master, aided the investigation in many ways both in furnishing the data for the table of ditches and by assistance in several instances in gaging the river and making seepage measurements. The many courtesies from ranchers all over the valley are also acknowledged.

Recommended for publication.

A. C. TRUE, *Director.*

Publication authorized.

JAMES WILSON,
Secretary of Agriculture.

WASHINGTON, D. C., *May 10, 1904.*

O